SEE, MAKE AND DO

AT THE SEASHORE

Pacific Edition

PAMELA HICKMAN

Illustrations by Twila Robar-DeCoste

Formac Publishing Company Limited
Halifax

Canadian Cataloguing in Publication Data

Hickman, Pamela
    At the seashore
    (See, make and do)
    Includes index.

ISBN 0-88780-404-7

1. Seashore biology — Pacific Coast — Juvenile literature. 2. Nature study — Activity programs. I. Robar-DeCoste, Twila L. (Twila Lee), 1956- .
II. Title. III. Series.

GL95.7.H55 1997 j577.5'1   C97-950113-X

Formac Publishing Company Limited
5502 Atlantic Street
Halifax, Nova Scotia B3H 1G4

Printed in Canada

Acknowledgements:
I wish to thank the following people for their assistance with my research for this book: Dr. John Ford, Andy Lamb, Catherine Po and Katherine Cook from the Vancouver Aquarium; Bob Pietrzak, Canadian Hydrographic Services; and Dr. Cheryl Lycette, N.D. for her advice on jellyfish stings. I would also like to acknowledge Dr. Merritt Gibson's excellent field guide, *Seashores.* Finally, I'd like to thank my friend Anne Kehoe for many enthusiastic seashore explorations and terrific campfires on the beach.

Dedication:
I would like to dedicate this book to the memory of David Threlford who shared his knowledge, enthusiasm and love of nature with so many.

Formac Publishing Company Limited acknowledges the support of the Canada Council for the Arts for our publishing program and the Nova Scotia Department of Education and Culture.

Distributed in the United States by
Seven Hills Book Distributors
49 Central Avenue, Cincinnati, Ohio 45202

Contents

Welcome to the seashore

Whether you are heading to your local beach or travelling to shores unknown, this book will give you lots of ideas of what to take, things to do, what to look for and where to look. In this section you'll find out about tides and learn how to make your own tide table. Discover the different life zones on a beach and make some great "buddies" to help you dig down for a closer look at shore life.

Heading out

An outing to the seashore is a great way to have fun and meet some of nature's most fascinating creatures. Whether you dig in the mud, poke into rock pools, paddle in the shallows or just lie on the beach, this book will help you and your family discover many new adventures that await you at the shore.

To ensure that your shoreline activities are safe for you, and for the creatures that live there, here are a few safety and conservation tips to keep in mind.

SAFETY TIPS

1. Children should explore in groups of two or more. Adult supervision is recommended.

2. Check a tide table so you know when high tide is expected.The best time to explore the shorelines below cliffs is after high tide, when the tide is going out.

3. Where possible, avoid walking on wet and algae-covered rocks — they are very slippery.

4. Take time to check out the shoreline before planning a long walk. Shoes are recommended for walking on barnacle-covered rocks. Tidal flats are often very muddy. As shoes and boots can get stuck in the mud, go barefoot.

5. Avoid rocky shorelines during high winds and storms. Large, powerful waves can appear without warning.

6. At certain times of year, shellfish are not safe to eat off the coast due to "red tide", a disease. Signs are usually posted for your safety, but you can check with the Department of Fisheries and Oceans to be sure.

CONSERVATION TIPS

1. Always take your garbage away with you. Pack an extra plastic

bag so you can pick up garbage left by others and make the shoreline a better place for everyone. Join up with friends and neighbours for the annual Beach clean-up and help clean up the garbage on a certain stretch of shoreline.

2. Handle living creatures carefully. When you are finished looking at them, always return them to where they were found.

3. If you find a nest, watch it from a safe distance and make sure that the adult animals are not frightened away. Do not touch the nest, eggs or young.

4. Enjoy the wildflowers where they grow, instead of picking them.

5. Do not climb on sand dunes. They are very sensitive to disturbance and can be easily damaged by trampling (see p. 11 for more on sand dunes).

6. If you move rocks to make a campfire, replace the rocks the way they were found.

Tide schedules

Check the local tourist bureau, a marina, dive or tackle shop, or newspaper for the current tide times in your area (the tide times are specific to each section of shoreline). For information on purchasing a set of tide tables for the west coast, contact the Hydrographic Chart Distribution Office in Sidney, B.C.. Or, turn the page and find out how to make your own tide table.

Tide talk

Take a walk on the ocean floor. Impossible? When you walk out on the mud flats, you are really walking on part of the ocean floor. Twice a day the ocean waters flow away from the shore; this is low tide. When the tide returns, it floods the mud with one to five metres of water, depending which part of the coast you are on. The mudflat that you walked is now deep underwater.

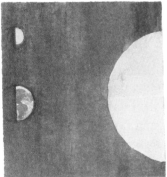

Neap Tides

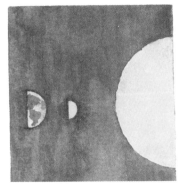

Spring Tides

TELLING THE TIDE

When you go down to the shore, look for the high tide mark along the sand or rocks. You can usually see a line of seaweed, driftwood and other things that have been left high and dry by the water. On large rocks, you may be able to see a difference in the kinds of plants and animals that are attached above and below the high tide mark.

What causes the tides? Tides are created by the gravitational pull of the Moon and Sun. Since the Moon is much closer to Earth than the Sun, its pull is stronger. Twice a month, when the Earth, Moon and Sun are in a straight line, the pull is strongest and the tides are highest. This occurs at the New Moon and the Full Moon and these tides are called spring tides. When the Moon, Earth and Sun make three points of a triangle, the tides are lowest, called neap tides. The height of tides is also affected by the shape of the ocean floor and the coastline.

Make a tide calendar

There are two regular tides per day on the northern Pacific shore of North America. It is a bit longer than six hours between high tide and low tide. You can calculate the tides for a shore near you with some simple math. Then you'll be able to predict high tide and low tide all summer, or even longer.

Remember, the time of high tide at one part of the shore can be very different from high tide at another place some distance along the coast.

You'll need:
• a calendar
• a current tide schedule
• two different coloured pencils

• Check your local newspaper for the high and low tide times at a shore near you. Mark these times, for the specific date they are given, on your calendar. Use one colour for high tides and the other for low tides.

• To figure out when high and low tide will be from one day to the next, add 50 minutes to the time of the tide. For example, if high tide comes at 6:00 a.m. and 6:25 p.m. on one day, it will come again at approximately 6:50 a.m. and 7:15 p.m. the next day. The same is true for low tides. By doing these simple calculations for each day on your calendar, you can make your own tide table for your section of the shore.

From sand to sea

As you explore the beach, you'll discover that the kinds of plants and animals change as you move from the dry sand dunes to the water. Each species does best in a certain area, or zone, of the beach. Some species prefer the driest zone of the dunes and upper beach (zone A), while others are adapted to the tides and can survive regular flooding (zone B). Certain plants and animals need to be underwater at all times (zone C). When you visit a beach, try to find the main beach zones illustrated below.

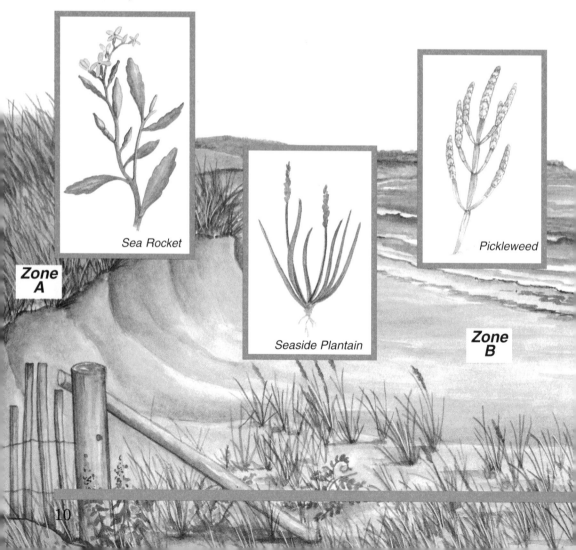

Sea Rocket

Zone
A

Seaside Plantain

Pickleweed

Zone
B

Help protect the dunes

When strong winds blow sand up the beach over a period of time, sand dunes are created. The wind keeps blowing the sand back from the shore until it is trapped by growing plants such as American Beach-grass. Its roots hold onto the sand and keep it from blowing away. As new plants grow, the dune becomes more stable and insects and other small animals move in. The entire habitat depends on keeping the sand from blowing away.

For this reason, it is very important that people stay off the dunes. Climbing on sand dunes damages the plants and exposes their roots so they can't hold onto the sand any more. When the sand is left unprotected, the wind will again blow it away. Other plants will be smothered by blowing sand and the dune habitat can be destroyed by a "blow out."

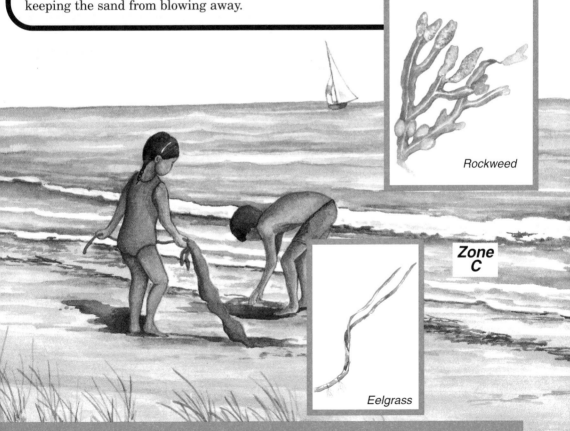

Rockweed

Zone C

Eelgrass

Gearing up

Before you head to the shore with your family or friends, pack your knapsack with some "beach buddies" and dress comfortably. Here are some tips on what to take.

BEACH BUDDIES
- snacks or lunch and fresh water
- sunscreen
- insect repellent (on calm days mosquitoes, blackflies, deerflies and horseflies can be a nuisance)
- first aid kit
- matches
- a towel
- an extra jacket or sweater
- binoculars
- field guides, such as birds, insects, seashells, seashores (see list on page 63)
- a plastic magnifying glass
- pad and pencil for making notes or sketches
- a camera
- something to dig with in the mud (see page 14)

- an underwater viewer (see page 15)
- a container for a temporary aquarium (see page 22)
- some plastic bags for collecting "treasures" and garbage
- picnic blanket
- a hat to keep off sun and insects
- loose-fitting pants that can be rolled up
- old sneakers for walking on rocks
- rain coat and rain hat
- knapsack
- bathing suit
- a watch

Beach "buddies"

A super scooper

Digging in the sand and mud is lots of fun. It's also a great way to find some of the creatures that live there. Here's a super scooper you can make out of an old plastic jug.

You'll need:

- a rinsed out 2-L juice jug with handle and lid (old bleach containers work well too)
- scissors
- sandpaper

1. With the handle facing up, ask an adult to cut out the lower part of the container, as shown.

2. Sand the cut edges until they are smooth.

3. Remember to fill in your holes when you are finished digging.

you can make

An underwater window

Take a peak under water with this easy-to-make viewer. You'll need:

- a rinsed out 1-L juice can
- a can opener
- waterproof tape (duct tape works well)
- clear plastic wrap

3. Stretch a piece of plastic wrap across one end. Tape the edges securely to the can.

1. Use the can opener to take both ends off of the can.

2. Tape the sharp edges of the can so you don't cut yourself.

4. To use the viewer, place the end with the plastic wrap in the water. Do not let the water come up over the open end. Look into the open end and see what's happening underwater.

On the beach

Find out where beaches come from and then discover what lives on and under the sandy shore. Dig up some of the hidden creatures in the mud so you can watch them more closely in a temporary home that you make. Comb the shores for "treasures" and make some wind chimes out of shells. Then come back to the shore at night and discover a whole new world of creatures that come out in the dark.

A beach is born

Have you ever wondered where all of the sand on a beach comes from? Follow the scenes below to see how a beach is born.

1. The grains of sand start out as part of much larger rocks, perhaps a long way from the shore.

2. Wind and water are constantly eroding, or wearing away, the rocks. The tiny particles are carried to the sea by rivers and are dumped on the ocean floor.

3. Currents and tides carry the sand up from the bottom and deposit it on the shore.

4. Waves move the sand higher onto the land, creating a beach and wind blows the sand farther up the beach, forming sand dunes.

Sand crafts

Sand is not only terrific to walk on and dig in, but you can also use it to make some great crafts. Try out some of these ideas.

- Draw a simple pattern on a piece of stiff paper or thin cardboard. Go over your lines with a thin layer of white glue and then sprinkle sand over the paper. The sand will stick to the glue and a sand picture will appear.

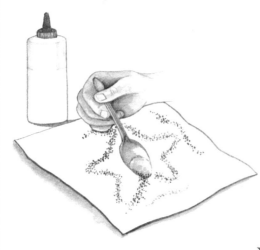

or whatever. By poking long sticks into the jar you can mix the colours in certain places and create patterns.

- Place some sand in two or three different plastic containers (yogurt or margarine containers work well). Thoroughly mix a different colour of powdered tempera paint into each container. Scoop layers of different coloured sand into a tall glass jar or vase to make a pretty anchor for dried flowers, candles,

- Mix some sand into liquid tempera paint to get a textured effect in your picture.

Sand tracks and

Spotted Sandpiper

Sand Sole

Sand Dollar

Pacific Razor Clam

When the
tide is out, take
a walk across the wet
sand. What kinds of markings
do you see in the sand? Some are
left by the receding water, while
others are signs of life at the
shore. On your next visit to the
beach, see how many of these
signs you can find.

As the tide goes out, the
backwash of the waves leaves
widely spaced ripples lying
parallel to the shore.

Bird tracks are easy to spot.
Narrow, three-toed tracks are
left by small shorebirds such as
sandpipers, sanderlings and
plovers. If you see large,
triangular
webbed feet, they are
probably gulls' imprints.
Deep, narrow holes near
shorebird tracks are made
when the birds poke their
beaks into the mud while
feeding. Shorebirds find their
food by feeling and smelling it.

Small, shallow depressions
in the sand are signs of
flounder. They come in
with the tide and take
"bites" out of the sand
in their search for
food in the mud.

trails

Lugworms live in the sand in the intertidal zone. With some practice you can find their tiny U-shaped burrows. The front opening is funnel-shaped and the narrower back opening is marked by tiny piles of castings, similar to the castings of earthworms.

If you notice water squirting out of groups of holes in the sand or mud as you approach, you have probably found a colony of Soft-shelled Clams. They are often found near a freshwater inlet. Pacific Razor Clams also squirt water and then quickly burrow deeper into the sand.

At low tide look for the outlines of Butter Clams or little neck clams buried just below the surface.

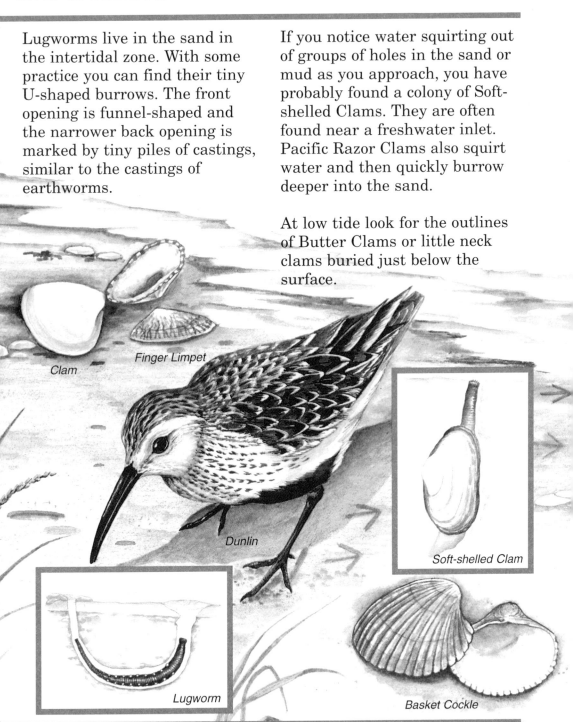

Clam

Finger Limpet

Dunlin

Soft-shelled Clam

Lugworm

Basket Cockle

Dig it

Once you find signs that life is lurking beneath your toes in the wet sand, start digging. Use your home-made scooper (see page 14) or a trowel and dig down at one side of a creature's hole in the sand. Work quickly since some of these animals, especially razor clams, can rapidly dig deeper. When you find something, carefully transfer it to the "mobile home" described below. Handle clam worms and sandworms carefully as they will nip.

Sandworm

Proboscis Worm

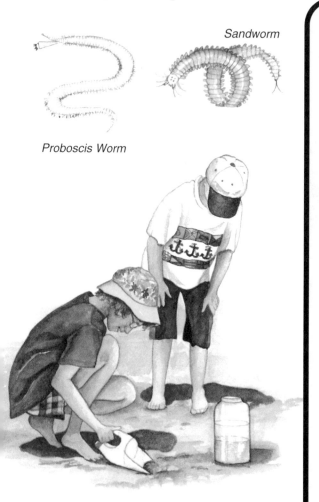

Make a "mobile home"

Set up this temporary aquarium for a closer look at how the animals move and feed.

You'll need:
- a rinsed out 4-L jar (a large condiment jar works well)
- some sand and sea water
- a scoop
- shoreline creatures dug out of intertidal mud
- a piece of cheesecloth or fine screening
- an elastic band

1. Scoop some sand into the bottom of the jar until it is about half full.

2. Add some animals that you have dug up in the intertidal zone, such as clams, mussels, worms, etc.

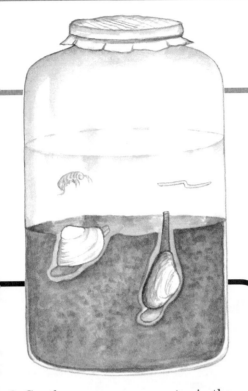

3. Gently pour some sea water in the jar until it is two thirds full.

4. Put a piece of cheesecloth or fine screening over the jar and secure it with an elastic band.

5. Place the jar in the shade and let the sand settle. Since the sand is covered with water, it is similar to high tide and the animals should be active. Watch for worms swimming around. Can you see a clam moving along on its fleshy foot? Try to find a clam's tube-like siphons poking out between its shells. A siphon sucks water and nutrients into the shell where the microscopic food is trapped and eaten. Waste water is discarded through another siphon.

6. When you have finished watching the creatures, return them to where you found them and fill in the holes you dug.

CLAMMING

At low tide from April to October you may see people bent over and digging on the mud flats. What are they digging for? The answer is butter clams or little neck clams. In addition to the sea ducks, crabs, sea stars and flatfish that naturally feed on clams, people also join in the harvest.

Whether you dig your own clams or buy them fresh (live) from a local market, try this simple method of preparing them at home or on a campfire.

1. Make sure the shells of fresh clams are tightly closed. Discard open shells since they mean that the clam is dead.

2. Scrub the shells under cold water to remove any sand.

3. Put the clams in a steamer over rapidly boiling water and cover the pan tightly.

4. Steam the clams until the shells open (about ten minutes).

5. Scoop the clams out of the shells and serve them with melted butter.

Shell collecting

It's fun to search the shore for special shells but instead of just collecting them, take a closer look at your "treasures." First of all, check to see if anybody's home. If there is an animal inside the shell, have a careful look at it, but let it go when you are finished. If the shell is empty it is safe to collect it.

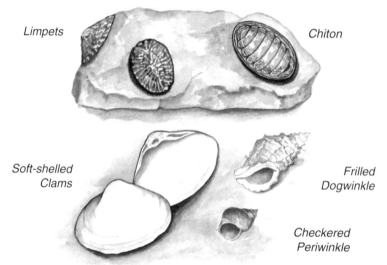

Limpets

Chiton

Soft-shelled Clams

Frilled Dogwinkle

Checkered Periwinkle

Shelled animals are divided into three main groups: chitons, bivalves and univalves. Check out the shells on these pages to see what kinds you have found.

- Chitons are easily identified by their eight overlapping plates. These small, flattened shells are usually found attached to rocks.

- Bivalves, meaning two shells, include the familiar clams, mussels and scallops. Their shells are hinged together and attached by several muscles. On the beach, most bivalves have become separated. Look closely to find the place on the shell where it was hinged. Can you see smooth, roundish areas on the inside of the shell? These are scars that show where the muscles were attached.

- Univalves, such as snails, periwinkles and limpets, have only one shell. Their shell may be in a coil or spiral (snails and periwinkles) or cap-shaped (limpets).

Shell bells

Make some pretty wind chimes for your garden or balcony. You'll need:

- thin, strong cord (available in sewing stores)
- scissors
- a large darning needle
- a variety of different shells with small holes in them
- two Popsicle sticks
- craft glue

1. Cut four equal lengths of cord, about 50 cm each.

2. Using the needle, thread approximately the same number of shells along each cord, tying a knot below each shell so it can't slip off. If there is no hole, try piercing thin shells with the needle.

3. Make a cross with the two Popsicle sticks and glue them together in the centre.

4. Tie one cord of shells to each arm of the sticks. The shells should hang close enough so that they will bump each other when blown by the wind. If the cords are lopsided, add more shells to even out the weight of each side.

5. Tie a piece of cord to the centre of the sticks and hang your wind chimes outside where they will catch the wind.

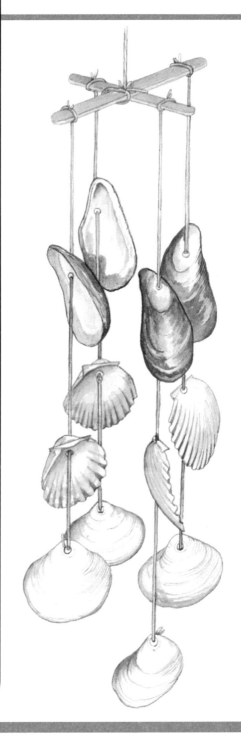

Beachcombing

When you're looking for shells on the beach, you will likely come across a number of other plant and animal remains that have been washed up by the wind and the waves. Look for these common beach "finds" and others. Use a field guide (see page 63) to help you identify what you find.

SAND DOLLARS

Look for dried, bleached sand dollars. The round, hard shells of these animals are often used as decorations. Notice the five "petals" radiating from a central hole on the top. Turn the shell over to find the mouth hole in the centre.

CRAB SHELLS

Bits and pieces of crab shells often turn up in the sand. As crabs grow, they shed their old shell and grow a new one. The old shells are discarded and wash up on beaches.

eye-like markings. Sponges look a bit like plants but they are really animals. If you have a hand lens, look closely at the pores covering the sponge. The sponge filters food out of the water that flows through the pores.

SKATE EGG CASES

If you find a small black, leathery "pouch" with long, curled spines, you've found the egg case of a skate. Also known as a mermaid's purse, the pouch contained the eggs and developing young of a skate, a fish with a long tail and wing-like fins that lives close to the ocean floor.

SEAWEED

Most seaweed grows attached to rocks underwater, but it often breaks off and is washed up on beaches. Try popping the balloon-like air bladders of Bladder Wrack (Rockweed) and Knotted Wrack, or wrap yourself in long ribbons of kelp. For more on seaweed, see page 36.

EYED FINGER-SPONGES

You may come across a piece of bleached spongy material with many finger-like branches and

The seashore at

Pack a late-night snack and head to the shore on an evening when the tide is going out. Dress warmly, bring an adult and a flashlight and you're ready to explore the shore at night. Many animals come out in the dark because they are hidden from their enemies. Other creatures survive better in the cooler, damp night air.

When you're walking or sitting on the sand, you may notice some reddish-brown or bluish-gray, jumping creatures about the size of a jelly bean. These are sandhoppers, sometimes called beach fleas. If you're quick, you can catch one in your hand —

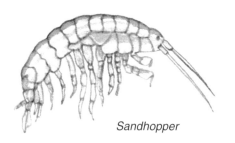
Sandhopper

they don't bite. Sandhoppers are tiny crustaceans, closely related to crabs and lobsters. During the day they burrow just under the sand in the upper beach to stay out of the hot sun. At night they venture out to feed on dead material left by the tide.

UP AND DOWN

The tiny floating plants and animals in the water, called plankton, move around at night. During the day, the plants float near the surface where the sun can reach them; at night they sink nearer the bottom. The animals move from the bottom to the top at night. Here they can feed more safely in the dark. Plankton is the most important life in the ocean because all other life depends on it. It is eaten by the tiny animals which are, in turn, eaten by larger and larger animals.

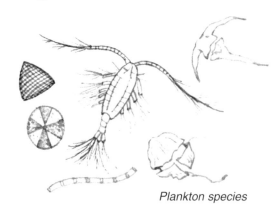
Plankton species

An underwater light

There's a lot happening in the shallow water at night, too. Tiny shrimps, small fish, crabs, worms, starfish and sea cucumbers are just some of the creatures swimming around in the dark. You can get a good look at them by making this simple underwater light.

You'll need:
- a flashlight
- a jar with a tight-fitting lid, large enough to hold the flashlight
- a heavy rock
- waterproof tape
- some string

1. Turn the flashlight on and place it in the jar with the light pointing down. Add the rock to the jar as a weight.

2. Put the lid on tightly and seal it with some waterproof tape.

3. Tie string around the jar and carefully lower it into the shallow water. Hold on to the other end of the string. What do you see?

Rocky shores

What clings to the rocks along the seashore? Plenty! Discover how the plants and animals differ between the high tide and low tide zones. Find a mini ocean on the rocks and explore the amazing life of a tidal pool. Try cooking with seaweed. Spend your day hunting for treasures and then share your discoveries around an evening campfire at the shore. Don't forget the marshmallows!

On the rocks

Where can you go to see a great collection of plant and animal life, from water-logged seaweed and mussels to high and dry lichens and snails? Just look at a large rock or rocky cliff at low tide. Can you see the different colours, almost like layers, on the rocks? Like the beach, a large rock can be divided into zones.

High tide mark

Intertidal zone

Low tide mark

Try to find where the high tide and low tide marks are. Between these marks is the intertidal zone. Take a hand lens and explore one of these rocks and discover the fascinating mini-world that clings to a rocky shore. Don't forget to check in the cracks and crevices, too.

BARNACLES UP CLOSE

About half way between the high and low tide marks, you'll find colonies of hard white barnacles clinging to the rocks. Take a close look at a barnacle. Notice the series of shell-like plates that are tightly closed at low tide. When the tide comes in, these plates open up and the barnacle's feathery legs reach out to sweep plankton into the animal's mouth. When a barnacle colony gets crowded, the shells of the oldest barnacles grow longer, and younger barnacles grow up beneath them. Eventually the old barnacles are cast off the rock. See if you can find some old and young barnacles.

Exploring a rock

When the tide rolls out, pools of water are left behind creating mini oceans on the rocks. A visit to a tidal rock pool is like a close-up view of life under the sea. Each pool will hold a different assortment of plants and animals, and all of them will offer you a glimpse of how the plants and animals depend on each other for survival.

What to look for

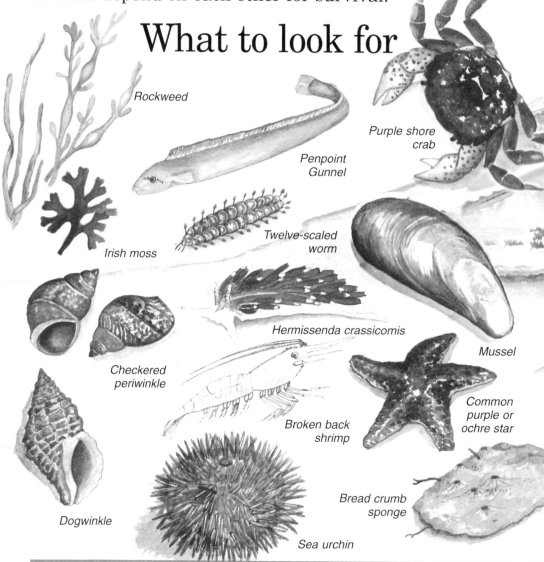

Rockweed

Penpoint Gunnel

Purple shore crab

Irish moss

Twelve-scaled worm

Hermissenda crassicomis

Mussel

Checkered periwinkle

Broken back shrimp

Common purple or ochre star

Dogwinkle

Bread crumb sponge

Sea urchin

pool

HIGH AND DRY?

When the tide is out, compare a tidal rock pool near the low tide mark with one near the high tide mark. Pools near the high tide mark are exposed to the hot sun for longer. Water evaporates more quickly from these pools, often leaving a white salt line around the rim of the pool. When water evaporates, the remaining water is saltier than the sea. The temperature of shallow pools can also rise very quickly. If you have a thermometer, compare the temperatures of the water in a rock pool near high tide and near low tide, to the ocean itself. The higher temperatures and saltiness in rock pools make it difficult for some creatures to survive.

Seaweed secrets

What do Sea Beard, Mermaid's Hair and Feather Bush have in common? They are all types of seaweeds. Seaweeds are large algae that grow in thick carpets along the coast.

They provide food and shelter for a huge variety of life including fish, periwinkles, sponges, sea cucumbers, and sea urchins. When you find a mass of seaweed, carefully check on and between the blades for the eggs and developing young of different marine animals. Notice the colour of the seaweed. There are three main groups: green, brown (including greenish-brown) and red (including purplish and nearly black). In general, you can find green seaweeds growing in the intertidal zone, brown ones growing near the low tide mark, and red seaweeds in the subtidal zone.

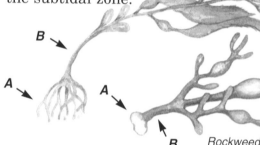

Rockweed

Seaweed for supper

Have you eaten any seaweed lately? If you've had commercial ice cream, pudding, frozen concentrates or chocolate milk, the answer is probably "yes." All of these things are made with carrageenin, a substance extracted from the seaweed called Irish Moss (Chondrus). Even your toothpaste might contain carrageenin! Here's another tasty way to eat seaweed. Irish Moss can be collected along rocky shores at low tide by raking.

The curly blades break off and the holdfasts remain on the rocks to grow new plants.

PARTS OF A SEAWEED

Unlike most land plants, seaweeds don't have roots, stems and leaves. Instead, they have a holdfast to attach them to a rock or other hard surface (A), a stipe which is similar to a leaf stalk (B), and a main body (C). When you examine a seaweed, try to find all its parts.

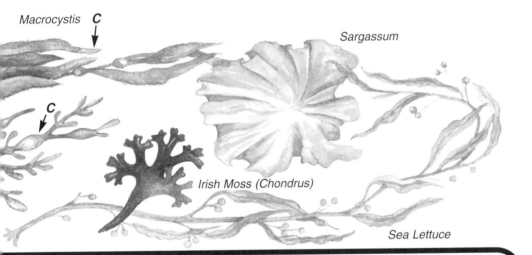

Macrocystis **C**

Sargassum

C

Irish Moss (Chondrus)

Sea Lettuce

You'll need:
- 1/3 cup of Irish Moss
- 4 cups of milk
- 1/4 teaspoon of salt
- 1 1/2 teaspoons of vanilla
- a double boiler
- a bowl
- sugar
- a kitchen sieve
- toppings, such as sugar, cream, fresh fruit or flavoured sauces

1. Rinse the Irish Moss thoroughly under cold water to get rid of sand and dried salt.

2. Place the seaweed and the milk in the top of the double boiler and boil for about half an hour until the milk is slightly thickened.

3. Add the salt and vanilla and mix well.

4. Pour the contents of the pan through the sieve into a bowl and chill. When the mixture has cooled it will look a bit like white jelly. Serve it with your favourite topping for dessert.

Living under the

The most important life in the ocean is so small that you can't see it. It's called plankton. Plankton is made up of tiny floating plants and animals that all other life feed on. Without plankton, there would be no other life in the ocean. If you have a microscope, collect a sample of water in a jar and place a drop of it on a glass slide.

Look at the water through the microscope. You'll see beautifully shaped green organisms called diatoms, as well as some tiny, moving animals. The tiny animals feed on the diatoms. Insects and other small animals feed on the plankton animals. They are in turn eaten by small fish and other small animals. Larger fish and birds eat the small animals, and mammals eat the large fish. This sequence of eating and being eaten is called a food chain. Since most plants and animals are eaten by more than one thing, several food chains overlap and form a food web. Look at the simple ocean food web below to see how different plants and animals depend on each other for survival.

COLLECT SOME CLINGERS

If you look beneath a wharf or on rocks below the low tide level,

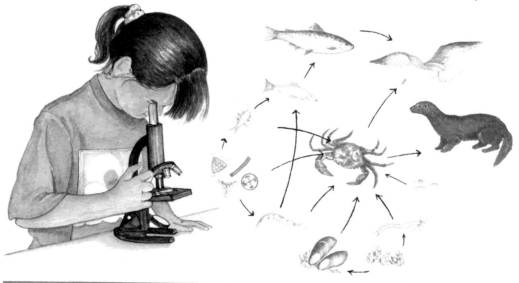

surface

you'll find a lot of different plants and animals clinging to these underwater surfaces. You can watch the amazing build-up of underwater plants and animals by "planting" a rock or weighted piece of wood in the water and checking it regularly over a week or two to see what has attached itself.

You'll need:
- a large rock and/or a weighted piece of wood
- a rope
- a magnifying glass
- a pencil and paper
- a camera (optional)

1. If you are using a piece of wood, tie a rock or other weight to it so it will sink in the water.

2. Tie a piece of strong rope to the rock or piece of wood and lower it into the water below the low tide line. Tie the other end of the rope to a wharf or other permanent anchor nearby, or to a floating buoy.

3. Every day, pull up your rock or piece of wood and check for life clinging to it. Use a magnifying glass to see the life close-up, especially the slimes that are the first to cling on. Make notes of what you see and, if you have a camera, take photographs.

4. At the end of your observations remove the rope and leave the rock or piece of wood in the water where the plants and animals can continue to grow.

Layers of life

Here's what you might find on your rock or piece of wood.

After a few hours: bacterial slime

After a day: plankton such as diatoms and protozoa

After two or three days: tiny animals such as hydroid and bryozoa.

After a week: larger animals such as barnacles and mussels, along with different algae.

Discovering dri

The smooth, weathered pieces of wood washed up on the shore are known as driftwood. Sometimes whole trees fall into the water at one location along the shore and they are carried far away by the currents and deposited on a different shore. The constant movement of the water against the wood smooths and shapes it before it is beached.

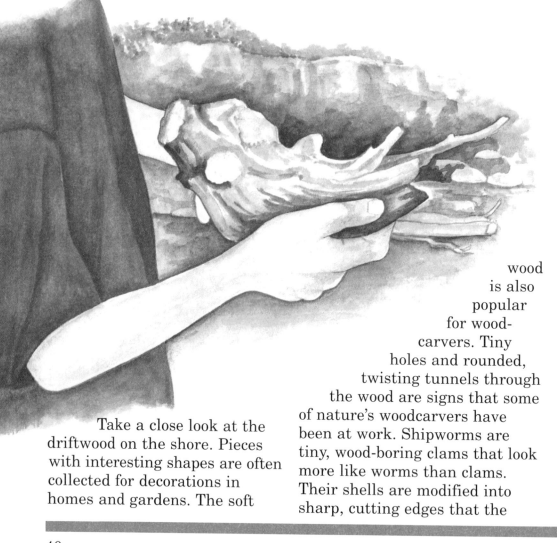

Take a close look at the driftwood on the shore. Pieces with interesting shapes are often collected for decorations in homes and gardens. The soft wood is also popular for woodcarvers. Tiny holes and rounded, twisting tunnels through the wood are signs that some of nature's woodcarvers have been at work. Shipworms are tiny, wood-boring clams that look more like worms than clams. Their shells are modified into sharp, cutting edges that the

ftwood

creatures use to "drill" into wood in the water. Unfortunately, shipworms can do a lot of damage to wharf pilings and other wooden structures in the water.

Make a driftwood pencil holder

Here's a simple craft you can make with a special piece of driftwood that you find at the shore.

You'll need:
- a small piece of driftwood at least 4 cm thick
- a drill and a 1 cm drill bit
- fine sandpaper

1. Sand your driftwood with find sandpaper to remove any splinters.

2. Ask an adult to drill several holes, about 3 cm apart and 2.5 cm deep where you want your pencils to go.

3. You can decorate your driftwood with paint, or glue on shells and other seashore finds. Use your imagination!

Campfires at the

If roasting hotdogs and marshmallows, singing songs and sitting out under the stars sounds like fun to you, then you'll love a campfire at the shore. Check the campfire tips and bring-along list below. Adult supervision is recommended.

CAMPFIRE TIPS

- build your campfire above the high tide mark. A ring of rocks around the fire pit will help contain the fire and keep young children from stepping into it.
- don't break branches off trees or shrubs for kindling.
- for cooking, keep the fire small since it will burn down more quickly, leaving hot coals perfect for roasting. Small fires are easier to put out at the end of the evening.
- when you are finished, let the fire burn right down. Smother the coals with water and sand until the fire is completely out.
- replace the rocks, remove all garbage and leave the shore as you found it.

Bring-alongs and sing-alongs

- food
- matches
- kindling and firewood, if driftwood isn't around
- a flashlight
- a penknife for sharpening roasting sticks
- a garbage bag
- container for carrying water to put out the fire

Sing-alongs and stories

There's something magical about telling stories and singing around a campfire. Try this oldtime favourite around your campfire.

Chorus

Fare - well to No - va Sco - tia, the sea - bound coast,

Let your moun - tains dark and drea - ry be,

For when I am far a - way on the bri - ny o - cean tossed,

Will you e - ver heave a sigh and a wish for me?

1 The sun was setting in the west,
The birds were singing on ev'ry tree,
All nature seemed inclined for rest,
But still there was no rest for me.

2 I grieve to leave my native land,
I grieve to leave my comrades all,
And my aged parents whom I always
held so dear,
And the bonny, bonny lass that I do
adore.

from *Folksongs of the Maritimes*, Kay Pottie and Vernon Ellis,
Formac Publishing 1992.

Some shoreline animals up close

Find out how to identify the gulls that come to your picnic and use our handy checklist to keep track of the long-legged shorebirds you see. Have you ever wondered where all the jellyfish come from in the summer? Become a nature detective and find clues about the animals that live at the shore and in nearby saltwater marshes. If you're going on a ferry, check out our guide to some common offshore wildlife. And now that you've spent time at the seashore and learned about the plants and animals that live there, gather your family and friends together to help protect the seashore.

Wings over water

Few birds are as easy to recognize as seagulls. Whether they are following fishing boats at sea, dropping mussels on the rocks to break them open, or trying to share your picnic, seagulls seem to collect near food.

GULLS

The most common gulls along the West Coast are Mew Gulls, Bonaparte's Gulls, Glaucous-winged Gulls and Herring Gulls (in winter). Their webbed feet help gulls swim and their long beaks are great for picking up and eating fish, shellfish, large insects, and dead animals. When you see a gull, try to decide what kind it is using this simple guide to adult gulls.

Mew Gull
- grey back and wings
- black wing tips
- small size

Glaucous-winged Gull
- grey back, wings and wing tips
- pink legs
- large size

Bonaparte's Gull
- black head in spring and summer
- black bill
- red legs

Herring Gull
- grey back and wings
- red spot on lower beak

Shorebirds

What has a long bill, long legs and is coloured to blend in with the sand and rock at the shore? If you said one of the twenty or so shorebirds found along the East Coast, then you're right.

Their long bills are adapted for probing into the mud to find worms and other small marine animals when the tide is out. Long legs help the birds to stay dry while walking at the edge of the water. Some shorebirds are found along the west coast all year round, while others are seen only during spring and fall migrations. The best time to view shorebirds is when the tide is high or coming in. This brings the birds closer to shore for easier viewing.

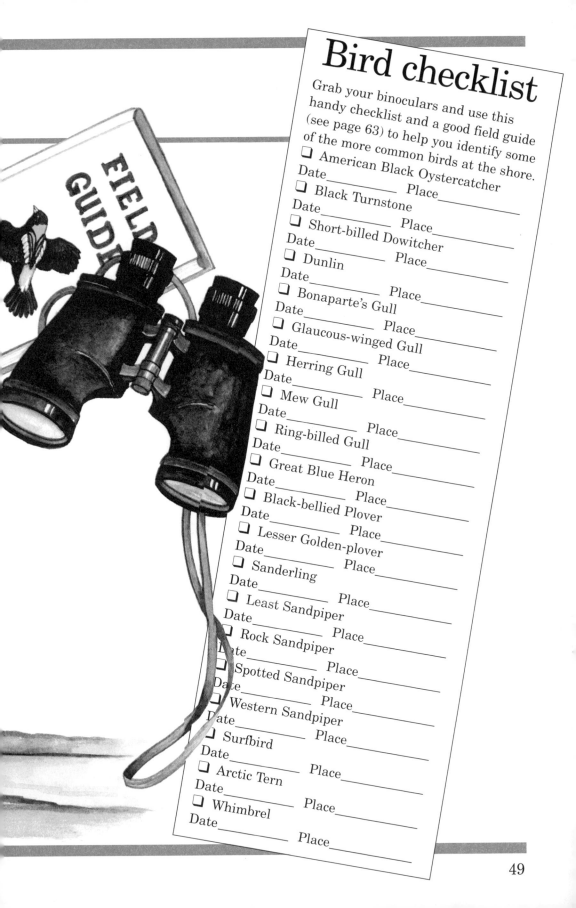

Bird checklist

Grab your binoculars and use this handy checklist and a good field guide (see page 63) to help you identify some of the more common birds at the shore.

❑ American Black Oystercatcher
Date_____ Place_____

❑ Black Turnstone
Date_____ Place_____

❑ Short-billed Dowitcher
Date_____ Place_____

❑ Dunlin
Date_____ Place_____

❑ Bonaparte's Gull
Date_____ Place_____

❑ Glaucous-winged Gull
Date_____ Place_____

❑ Herring Gull
Date_____ Place_____

❑ Mew Gull
Date_____ Place_____

❑ Ring-billed Gull
Date_____ Place_____

❑ Great Blue Heron
Date_____ Place_____

❑ Black-bellied Plover
Date_____ Place_____

❑ Lesser Golden-plover
Date_____ Place_____

❑ Sanderling
Date_____ Place_____

❑ Least Sandpiper
Date_____ Place_____

❑ Rock Sandpiper
Date_____ Place_____

❑ Spotted Sandpiper
Date_____ Place_____

❑ Western Sandpiper
Date_____ Place_____

❑ Surfbird
Date_____ Place_____

❑ Arctic Tern
Date_____ Place_____

❑ Whimbrel
Date_____ Place_____

Jellyfish

Jellyfish get their name from the strong, jelly-like substance that forms their body. The sac-like body is shaped like an umbrella with a ring of tentacles around the edge. In the centre of the bottom is a mouth that leads to a large stomach. Around the mouth are "arms" that help put food, such as small fish, plankton and crustaceans, into the mouth. Jellyfish are in turn eaten by some fish and sea turtles.

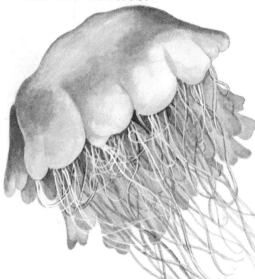

full. Now the jellyfish quickly contracts the muscles again and forces the water out with such power that the jellyfish shoots forward, like a rocket.

A DOUBLE LIFE

The life cycle of a jellyfish includes a non-swimming stage when it is called a polyp and lives attached to a rock, wharf or other solid object underwater. Later it becomes a free- swimming umbrella-shaped jellyfish that you see in the water. In early spring thousands of tiny jellyfish break off from the colonies of polyps and spend the summer drifting with the currents. By mid-summer they have grown into large jellyfish that may appear in huge numbers near the shore. During August and September the jellyfish lay their eggs and the young hatch and attach themselves underwater to spend winter in the

ON THE MOVE

Jellyfish float on the open water and are carried a long way by the currents and the wind, but they can also swim. When a jellyfish wants to move, it contracts the muscles near the outer rim of the umbrella shape so the umbrella looks like it is closing up. When it relaxes the muscles, water rushes into the animal's stomach until it is

non-swimming stage. Soon after egg-laying, the adult jellyfish die.

STRINGS OF STINGS

The long tentacles of a jellyfish contain stinging cells that are used to capture food and for self-defence. Stinging cells contain

poison to paralyse prey, and some can shoot out masses of tiny threads to tangle prey. Both types of jellyfish commonly found off the West Coast in late summer and early fall — Moon Jellyfish and Lion's Mane or Red Jellyfish — should be avoided, but the sting of the Lion's Mane is the most severe. If you happen to find a jellyfish washed up on the beach, do not touch it because its tentacles can still sting several hours after the animal has died.

Home-made remedies for jellyfish stings

If stung, try one of these home-made remedies to help relieve the pain and rash.

- Put vinegar or a cola drink on the sting to relieve pain.

- Make a herbal tea by adding a handful of leaves and flowers from St. John's Wort (a common roadside wildflower) and Stinging Nettle (use gloves to pick this) to two cups of boiling water. Soak a cloth in the water and bathe the stings. You can make the tea ahead of time and carry it to the beach in a thermos.

- A cloth soaked in hot water can help relieve the pain of a sting.

What a crab!

When you grow bigger, you grow out of your clothes and need new ones. But imagine if you also had to get a new house! That's what hermit crabs have to do. Instead of just living in their own shell like most crabs, hermit crabs live inside the abandoned shell of a snail.

When the crabs grow, their house gets too small and they have to find a bigger snail shell to call home. Look for hermit crabs scurrying over the sand and mud near the water line or along the bottom in shallow water. You may also find them in tide pools, especially near the base of large rocks. When they are frightened, they pull their legs into the shell and cover the opening with the large pincers attached to their front legs.

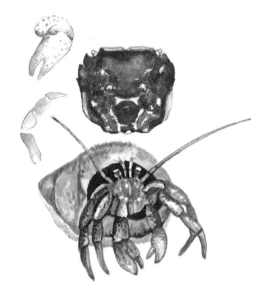

EMPTY SHELLS

Along the beach you may find the empty shells of some other common crabs — Red Rock Crabs and Green or Purple Shore Crabs. As these crabs grow, their hard shells split along the back where the body and tail are joined and the crab crawls out. This is called moulting. Beneath the old shell, the crab grows a new one that hardens in a few weeks. Before the new shell hardens, crabs have no protection from predators so they stay hidden in seaweed to avoid hungry fish, herons, gulls and diving ducks. To find young crabs, check under pieces of wood or rocks on sandy and rocky shores and in tide pools. Adults are often partly buried in the sand, with their large pincers near the surface ready to grasp worms, mussels, sea urchins or dead fish.

Use a field guide to identify these other common crabs: Kelp Crab and Decorator Crab.

Hold a crab race

If you
find a crab
on the shore, watch
how it moves. All crabs run
sideways. Try to run like a crab and then
hold races with your family and friends. First, lie down
on your back with your arms at your side. Now, push yourself up with your feet
and hands so that your body is off the ground. Try moving sideways. Who is the
fastest crab in your group?

Saltmarshes

In sheltered bays and other shoreline areas that are protected from rough waves, you may find saltwater marshes. These rich ecosystems provide food and shelter for an amazing assortment of plants and animals.

A number of flowering plants including grasses, sedges and rushes grow in saltwater marshes. Their roots trap sediment from the water and bind it together so that it cannot be washed away by the tides. Many of these plants are specially adapted to the salty water. They have thick, fleshy stems or leaves

that store fresh water and can get rid of the salt that is absorbed through their roots.

MARSH ZONES

Like other shoreline areas, saltwater marshes are divided into zones — high marsh and low marsh — depending on how much they are flooded during tides. Plants typical of the high marsh include Marsh Hay, Sea Lavender, Seaside Plantain, Milkwort, Arrow Grass and a variety of grasses, reeds, rushes and sedges.

The low marsh is home to cordgrass, some algae, Glasswort, Eelgrass, Orach and Seaside Sand Spurry. Many of these plants are eaten by ducks and geese, and they provide shelter and food for a variety of crabs, insects, spiders, Soft-shelled Clams, snails and Ribbed Mussels. Muskrats also feed on the stems and roots of marsh plants and use the plants to make their dome-shaped homes in the marsh.

Crafty plants

Two saltmarsh plants are favourites for crafts. Sweetgrass, which grows at the upper edge of the high marsh, has a beautiful fragrance and is traditionally used in basket-making and other crafts. Sea lavender also grows in the high

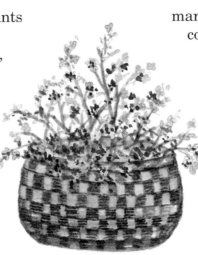

marsh. Its dense colonies of small, bluish flowers can be collected in the fall and hung upside down to dry. They are lovely additions to dried flower arrangements, wreaths and other decorations.

Become a nature

When you go for a walk at the shore, you may not see many of the animals that live there or visit at certain times of the day or night. But you can discover what is around by being a nature detective and looking for the clues they leave.

For instance, raccoons and mink may visit the water's edge at night for food — clams, mussels, worms and other creatures that are left behind when the tide goes out. Watch for their tracks in the wet sand before the tide comes in.

Muskrats may live nearby in a shallow saltwater marsh. They can be identified by tracks, but even more easily, by their dome-shaped houses made out of dead marsh vegetation and mud.

Many birds feed and nest at the shore. Look for the large nests of sticks built by Osprey, often in dead trees near the shore. Great Blue Herons also build nests of sticks and grass in the tops of

detective

large trees. Since these birds nest in colonies, you will see several nests in a single tree. Northern Rough-winged Swallows and Belted Kingfishers prefer nesting in holes near the tops of sandy cliffs. Spotted Sandpiper nests are simply depressions on the ground lined with leaves and grass. They are usually hidden in the grass or under a shrub in the upper beach area and are active in late May and June. All bird nests should be watched from a safe distance so the birds are not disturbed. Since many birds return to the same nest year after year, nests should not be removed or collected.

Animals ahoy!

If you take a ferry trip to the islands off the west coast, or you sit on a rocky shore and watch out over the water, you may see the sleek, bobbing head of a seal or a spout from one of the magnificent whales that live in the ocean. Observing whales, seals, porpoises, dolphins and seabirds is a great hobby for boaters and landlubbers alike.

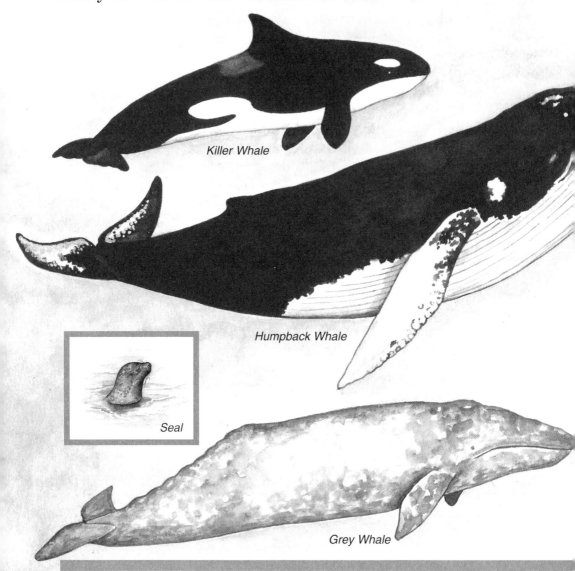

Killer Whale

Humpback Whale

Seal

Grey Whale

Harbour Porpoise

Pacific White-sided Dolphin

Best times for viewing

Use this checklist of some common wildlife seen from the shore and a field guide to help you with identification.

Animal	Spring	Summer	Fall	Winter
Pacific White-sided Dolphin	X			
California Sea Lion	X		X	
Steller Sea Lion	X	X	X	
Harbour Porpoise		X	X	X
Harbour Seal		X	X	X
Grey Whale		X	X	
Humpback Whale		X		
Killer Whale	X		X	
Minke Whale	X	X	X	
Double-crested Cormorant	X	X	X	X
Pelagic Cormorant	X	X	X	
Barrow's Goldeneye	X	X	X	X
Bufflehead	X	X	X	X
Canada Goose		X	X	X
Common Loon	X			X
Red-throated Loon	X	X	X	X
Harlequin Duck	X	X	X	X
Common Merganser	X	X	X	X
Red-breasted Merganser	X	X	X	X
Oldsquaw		X	X	X
Black Scoter		X		X
Surf Scoter	X			X
White-winged Scoter		X	X	X

WHALE WATCHING

Whales are most commonly seen in the spring and fall as they pass by our shores during migration between their summer feeding grounds in the north and their winter birthing grounds in the south. Look for their spouts, dorsal fins, and tail fins as they dive. Large flocks of seabirds feeding out at sea may be a clue to nearby whales who are also feeding on the fish. Some special whale watching sites in British Columbia include Tofino, Ucluelet, Victoria and Telegraph Cove. Whales may also be sighted from any of the ferries running between the mainland and Vancouver Island as well as other outlying islands.

S.O.S. Save our

Throughout this book you have discovered the amazing variety of plants and animals that call the seashore home. These plants and animals depend on each other for survival, but none can survive without their habitat

Whether it's a rocky shore, sandy beach, mudflat or saltmarsh, the destruction of a habitat leads to the loss of the species that live there. Some of the major threats to shoreline species include pollution from oil spills, toxic chemicals and garbage, and motor vehicles. Several endangered species, including the Leatherback Turtle and Northern Right Whale depend on healthy shorelines and waters for their survival. You can help protect the shoreline and its wildlife by following the Conservation Tips on page 7. Spread the word to your family and friends that the shoreline is a great place to visit and tell them how they can help protect

Leatherback turtle

it, too. You can also volunteer your time or money to help organizations that are working to protect the shoreline and its species. Contact the local naturalists' group, the Burrard Inlet Environmental Action Program, or the Marine Life Sanctuaries Society of British Columbia for information on groups that are involved in shoreline conservation work.

Right Whale

KILLER WHALE ADOPTION PROGRAM

Here's your chance to help with research that may protect present and future populations of Killer Whales, or Orcas. Over 450 Killer Whales inhabit the waters along British Columbia's coast. Marine biologists are studying these fantastic mammals to learn more about their needs for survival and the effects of human activities on their success. Problems such as pollution, overfishing and boating may all be threats to the whales. Through the Killer Whale Adoption Program you can adopt a whale of your choice and help to pay for this important work. The program is headed by well known marine mammal scientist, Dr. John Ford of the Vancouver Aquarium.

When you adopt a Killer Whale, you get:

- an official adoption certificate
- a biographical sketch and ID photo of your whale
- a cassette tape of Killer Whale sounds
- a copy of the program's annual newsletter, *The Blackfish Sounder*, so you know the latest news on Killer Whale research

For an application form write to: Killer Whale Adoption Program, Vancouver Aquarium, P.O. Box 3232, Vancouver, B.C., Canada, V6B 3X8.

Killer Whale

West Coast Parks in Canada and Northwestern United States

When your family is travelling along the west coast in Canada and the U.S.A. check out some of these national, provincial or state parks and recreational areas. All of them are located along the shore where you can explore their beaches and try out some of the activities in this book.

British Columbia
Information:
1-800-663-6000

National Parks
Pacific Rim
South Moresby Gwaii Haanas

Provincial Parks and Recreational Areas
Anthony Island
Brooks Peninsula Rec. Area
Cape Scott
Carmanah Pacific
Desolation Sound Marine
Echo Bay Marine
Fillongley
Fiordland Rec. Area
French Beach
Gibson Marine
Hakai Rec. Area
Loss Creek
Maquinna
Matheson Lake
McDonald
Miracle Beach
Naikoon
Oliver Lake
Porteau Cove
Rathtrevor Beach
Roberts Creek
Rugged Point Marine
Saltery Bay
  Smelt Bay

Washington State

National Parks
Olympic
San Juan Island National Wildlife Refuge

State Parks and Recreational Areas
Bay View
Birch Bay
Blake Island Marine
Carmano Island
Deception Pass
Fay Bainbridge
Fort Casey
Fort Elsey
Fort Flagler
Fort Warden
Grayland Beach
Illahee
Jarrell Cove
Joemma Beach
Kitsap memorail
Kopachuk
Lighthouse Marine
Lime Kiln
Manchester
Moran
Oak Harbor Beach
Ocean City
Odlin County
Pacific Beach
Penrose Point
Phil Simon
Point Defiance
Saltwater
San Juan County
Scenic Beach
Sequim Bay
Semiahmoo County
South Whidbey
Spencer Spit
Sucia Island
Tolmie
Triton Cover
Twin Harbors

Suggested Field Guides

Washington
Wenberg

Oregon

National Parks
Oregon Dunes National
  Recreation Area

*State Parks and
  Recreational Areas*
Bastendorff Beach
Beachside
Beverly Beach
Bullards Beach
Cape Arago
Cape Kiwanda
Cape Lookout
Devil's Elbow
Ecola
Fogarty Creek
Fort Stevens
Harris Beach
Neptune
Ona Beach
Oswald West
Samuel H. Boardman
Seal Rock Wayside
South Beach
Sunset Bay
Umpqua Lighthouse

Northern California

*National Parks and
  Recreational Areas*
Golden Gate National Rec.
  Area
Kings Range National
  Conservation Area
Point Reyes National
  Seashore
Redwood National Park

*State Parks, Beaches and
  Recreational Areas*
Anchor Bay Campground
Angel Island State Park
Baker Beach
Bodega Dunes
  Campground
Casper Headlands State
  Beach and Rec. Area
China Beach
China Camp State Park
Clam Beach County Park
Del Norte Coast Redwoods
  State Park
Dillon Beach
Doran Park
Fort Ross State Historic
  Park
Humboldt Lagoons State
  Park
Little River State Beach
MacKerricher State Park
Manchester State Beach
Muir Beach
Patricks Point State Park
Pelican State Beach
Prairie Creek Redwoods
  State Park
Salt Point State Park
San Elijo State Beach
Seal Rocks Beach
Sinkyone Wilderness State
  Park
Sonoma Coast State Beach
Standish-Hickey State Rec.
  Area
Stillwater Cove Regional
  Park
Stinson State Beach
Tomales Bay State Park
Trinidad State Beach
Westport-Union Landing
  State Beach

Field guides can help you
identify some of the
things you will find
during your seashore
adventures. Look for
these books in your local
library or bookstore.

*A Field Guide to
Pacific Coast Shells* by
Morris. Boston: Houghton,
Mifflin Co., 1989.

*A Field Guide to
Seashores* by J. Kricher.
Boston: Houghton,
Mifflin, 1992.

*A Field Guide to
Western Birds* by Roger
Tory Peterson. Boston:
Houghton, Mifflin Co.

*Beachwalker: Sea Life
of the West Coast* by
Stefani Hewlett Paine.
Vancouver: Douglas &
McIntyre, 1992.

*Seashore Life on Rocky
Coasts* by Judith Connor.
California: Monterey Bay
Aquarium, 1993.

*Seashores. A guide to
Animals and Plants
along the Beaches,* by
H.S. Zim and L. Ingle.
Golden Press, New York,
1989.

Index